Igneous Rocks

by Grace Hansen

Abdo Kids Jumbo is an Imprint of Abdo Kids
abdobooks.com

abdobooks.com

Published by Abdo Kids, a division of ABDO, P.O. Box 398166, Minneapolis, Minnesota 55439.

Abdo Kids Jumbo™ is a trademark and logo of Abdo Kids.

Printed in the United States of America, North Mankato, Minnesota.

052019

092019

Photo Credits: iStock, Science Source, Shutterstock

Production Contributors: Teddy Borth, Jennie Forsberg, Grace Hansen
Design Contributors: Dorothy Toth, Pakou Moua

Library of Congress Control Number: 2018963350

Publisher's Cataloging-in-Publication Data

Names: Hansen, Grace, author.

Title: Igneous rocks / by Grace Hansen.

Description: Minneapolis, Minnesota : Abdo Kids, 2020 | Series: Geology rocks! set 2 | Includes online resources and index.

Identifiers: ISBN 9781532185571 (lib. bdg.) | ISBN 9781532186554 (ebook) | ISBN 9781532187049 (Read-to-me ebook)

Subjects: LCSH: Igneous rocks--Juvenile literature. | Rocks--Identification--Juvenile literature. | Geology--Juvenile literature.

Classification: DDC 552.1--dc23

Table of Contents

Igneous Rocks

Earth is filled with rocks big and small. All rocks are made up of **minerals**.

Igneous rocks are some of the oldest rocks on the planet. They make up around 95% of Earth's **crust**.

Earth's **crust** has cracks in it, making many large pieces. These pieces are called **tectonic plates**. Most new igneous rocks form along the edges of these plates.

Below the **crust** is the **mantle**.
In the lower part of the mantle, temperatures get very hot.
It is so hot, that rocks melt!
This is where the formation of igneous rocks begins.

Melted rock is called magma. Magma rises from the **mantle** toward the **crust**. It seeps into cracks, cools, and becomes solid.

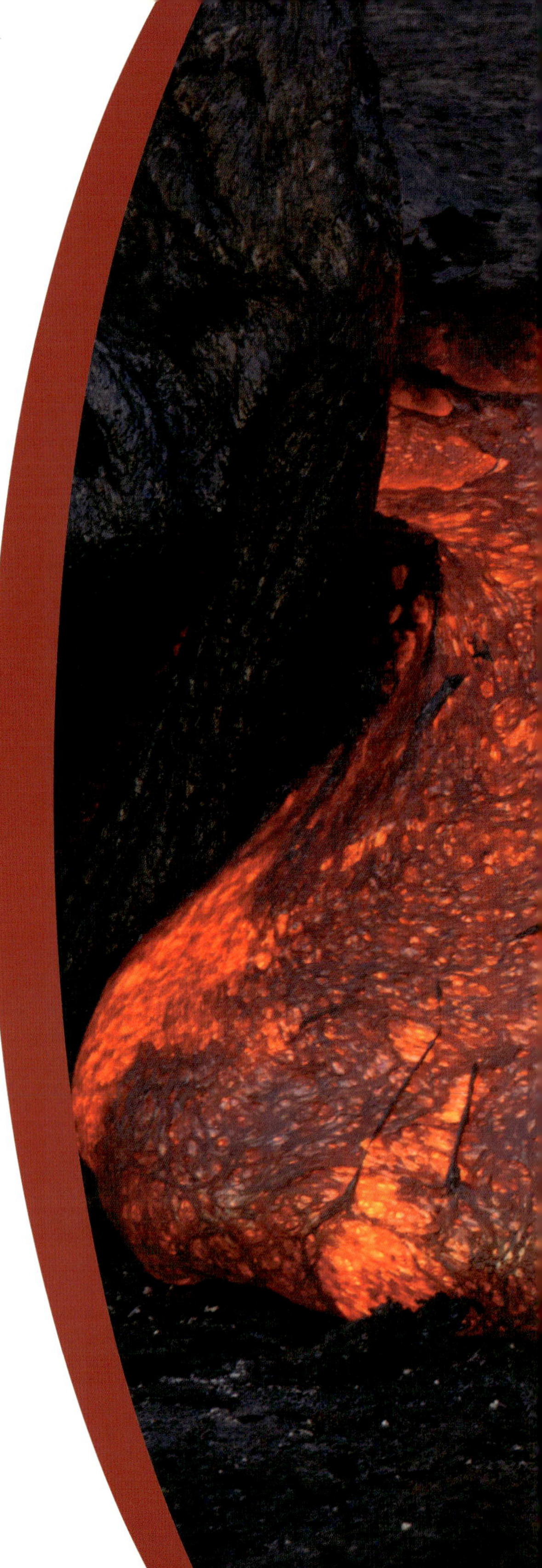

Sometimes magma comes to Earth's surface. Magma above ground is called lava. Lava also cools and hardens. Hardened lava and magma form igneous rock.

The **minerals** in the magma form crystals as they cool. The slower the magma cools, the larger the crystals are!

Common Minerals

Common **minerals** in igneous rock include feldspar and quartz. Granite is an igneous rock. It is made up of at least 25% quartz. Granite is strong and often used in building.

The Rock Cycle

Over the course of many years, igneous rock breaks down. It will form into **sedimentary** and then **metamorphic rock**. This process is called the rock cycle.

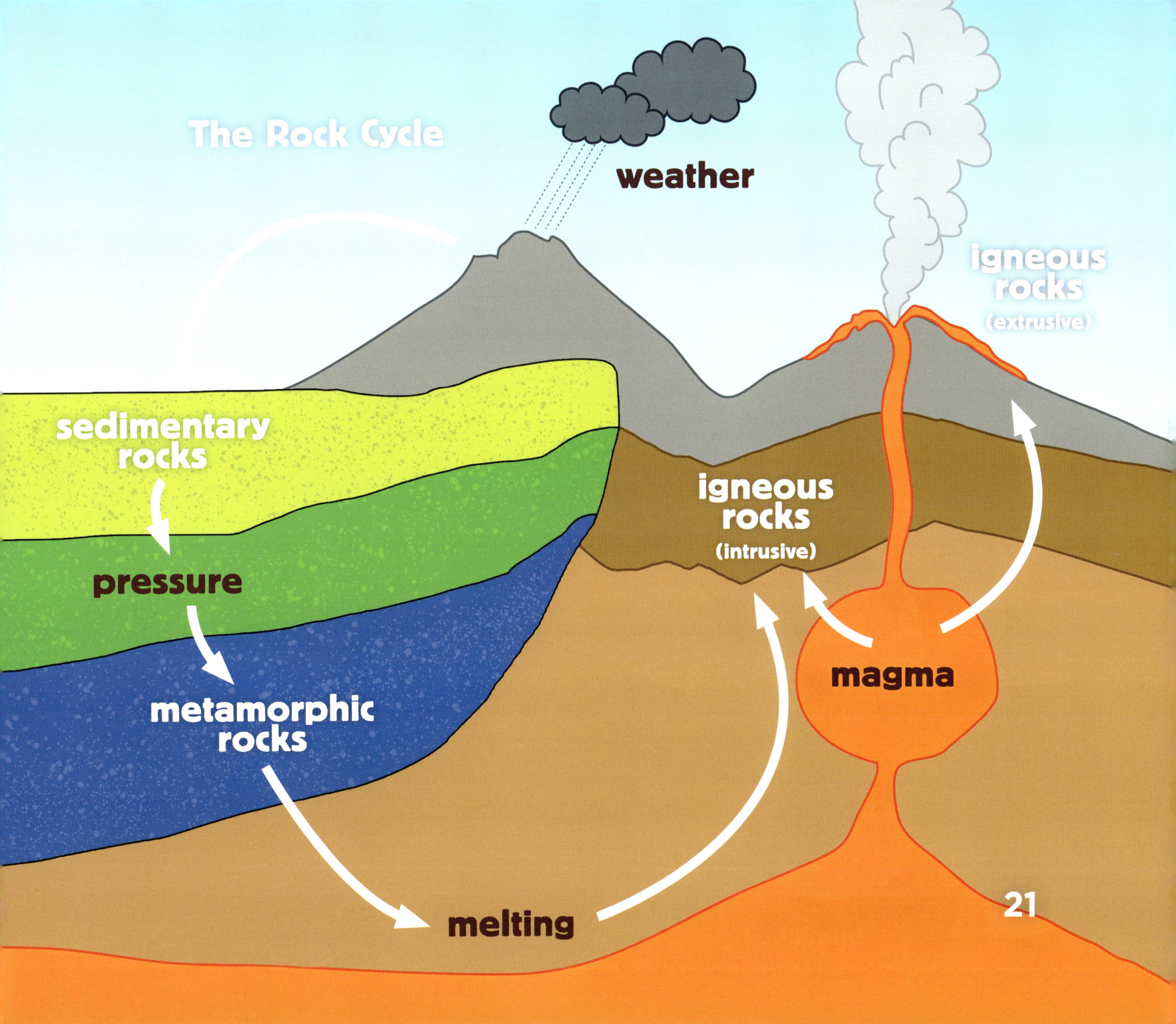
The Rock Cycle
weather
igneous
rocks
(extrusive)
sedimentary
rocks
pressure
metamorphic
rocks
igneous
rocks
(intrusive)
magma
melting

Extrusive & Intrusive Igneous Rocks

extrusive

Form above ground and cool quickly forming few or no crystals.

basalt

obsidian

pumice

intrusive

Form below ground and cool slowly, allowing large crystals to form.

diorite

granite

peridotite

Glossary

crust – the outer layer of Earth.

mantle – the layer of Earth that lies between the crust and core.

metamorphic rock – a rock type that arises from the transformation of existing rock by things like heat and pressure.

mineral – a substance, like gold, silver, and iron, formed in the earth that is not of an animal or plant.

sedimentary rock – rock that is formed by the accumulation and cementation of mineral or organic particles.

tectonic plate – any of the several segments of Earth's crust that move in relation to one another causing things like volcanic eruptions.

Index